SECRETS *of the* SPAN

LIONS GATE RENEWED

LILIA D'ACRES

 FriesenPress

Suite 300 - 990 Fort St
Victoria, BC, V8V 3K2
Canada

www.friesenpress.com

Copyright © 2016 Lilia D'Acres
First Edition — 2016

Edited by Bryn D'Acres

All rights reserved.

No part of this publication may be reproduced in any form, or by any means, electronic or mechanical, including photocopying, recording, or any information browsing, storage, or retrieval system, without permission in writing from FriesenPress.

ISBN
978-1-4602-7633-4 (Hardcover)
978-1-4602-7634-1 (Paperback)
978-1-4602-7635-8 (eBook)

1. TECHNOLOGY & ENGINEERING, CIVIL, BRIDGES

Distributed to the trade by The Ingram Book Company

TABLE OF CONTENTS

Prologue 1
Renovation 1972-1999 5
   Guinness Money 5
   The Seventies 6
   The Eighties 7
   The Nineties 9
Renovation 1999-2002 13
   The Contract 13
   Earthquake Design 14
   Roadway Deck 15
   Hangers 16
   The Sidewalks 17
   Winds 18
   Canron 18
   American Bridge/Surespan A Joint Venture . . . 20
   Mediated Settlement 23
   Secret Power 24
   Safety 26
   Monument to the Ironworkers 28
   Project Patrol 30
   Government 32
   Indian Land Claims 33
   Guardians of the Gate 34
Acknowledgements 35
About the Author 37
Sources 39

RAVEN DANCER (Alano Edzerza)

To Bryn

*Here's fine revolution,
an we had the trick to see it — Hamlet*

Thanks with love

SECRETS *of the* SPAN

LIONS GATE RENEWED

This island's mine, by Sycorax my mother,
Which thou takest from me. When thou camest first,
Thou strokedst me and madest much of me, wouldst give me
Water with berries in't, and teach me how
To name the bigger light, and how the less,
That burn by day and night: and then I loved thee
And show'd thee all the qualities o' the isle,
The fresh springs, brine-pits, barren place and fertile:
Cursed be I that did so! All the charms
Of Sycorax, toads, beetles, bats, light on you!
For I am all the subjects that you have,
Which first was mine own king: and here you sty me
In this hard rock, whiles you do keep from me
The rest o' the island.

The Tempest

PROLOGUE

COPPER WOLF
(Alano Edzerza)

Lions Gate Bridge holds high its secrets of engineering, drawn from two incarnations that only the bridge engineers are party to but that the world applauds. A.J.T. Taylor's genius in the first design gave expanse to possibility for all participants seeking greater success in capital and artistic ventures. Taylor himself would take his talent to the Empire State Building in New York.

Peter Buckland raised the bar higher with the incomparable redesign of the bridge. Buckland & Taylor shrugged off government opposition to their four-lane bridge proposal to complete an award-winning rehabilitation and seismic retrofit that expanded horizons for all. Their leading-edge designs have transfigured the world's waterways.

In both endeavours, the bridge ensured life for itself and its participants, giving dreams to players big and small. For the First Nations rescinded from participation, it cast a dark light on the world standards it has otherwise enjoyed. Built on Capilano Reserve No. 5 cut-off lands by government order, it took away homes of hundreds of villagers for token compensation. A court decision that returned remnants of the lands on either side of the bridge right-of-way, also forced First Nations families to forfeit ancestral land forever.

This dire act was not new to the First Nations people. By provincial law dating from 1913, First Nations were prohibited from living beside white people. Violations resulted in burning of homes and total destruction of gardens. Xwayxway (Whoi Whoi), their thousand-acre ancestral home—including five villages shared by the Musqueam, Skwxwu7mesh (Squamish) and Tsleil Waututh (Burrard) Nations—was destroyed. The losses for First Nations never stopped. Xwayxway, their beloved homeland, became Stanley Park, their access to it diminished. For all others, it became the beautiful conduit to the bridge and their dreams. Jurisdiction over Stanley Park is subject to land claims by the Skwxwu7mesh, Tsleil Waututh and Musqueam Nations.

The diminishment of First Nations had begun at first contact with whites. Prior to contact, they had numbered one hundred and twenty thousand. After contact, seven hundred. As more and more lands were taken away, fewer people could survive. Only three to five elders who had homes in Xwayxway remain on their land.

Emily Baker, former resident and wife of the late Chief Simon Baker, suggested in the advent of the 2010 Vancouver Olympics that Xwayxway be returned in name at least. When proposed to the federal government, it was declined without discussion. This too was not new.

Parliamentary recognition of Reserves in 1935 had rendered legal action almost impossible. Prior to 1960, the Federal Government declined First Nations the right to vote in municipal, provincial and federal government elections. They had no right to legal counsel and were thrown in jail for wearing regalia. Attempts to right the wrongs regarding personal livelihood and welfare were blocked. Since then, the statute of limitations limits legal action. 'There has been very little chance to deal with the wrongs. Never has the government come to us', says Chief Bill Williams. 'We have always had to express an interest first.' And apologies are words only when actions do not follow.

Having lost their lands to bridge building, they applied for jobs promised—jobs that never materialized. Getting the right to build a new bridge was ultimately possible under a new NDP government—their one chance at the bridge game. Partnered with SNC Lavalin, they submitted two designs: a twin-bridge concept that would preserve the existing bridge, and a two-tiered six-lane design. Neither made the final cut. Three First Nations painters were hired.

The King and Queen of England drove across the Lions Gate Bridge in the 1939 opening, not stopping for the reception by First Nations waiting in ceremonial dress. It would be remembered as the most personal of inhuman acts in bridge history. For the remainder of the century and beyond, the Skwxwu7mesh Nation would be relegated to Reserve land beneath the bridge, having to adapt to the unremitting rampage above, benefits not falling their way. They had not been honoured as participants.

Wealth wrought by the Lions Gate Bridge, for those who had been so honoured, is unfathomable. 'Creating our own wealth is what we've learned to do', says Chief Bill Williams. 'We're doing it deliberately, so there is no need to go to government.'

The bridge has sustained its landmark status and perhaps is standing stronger in new climes. Just beneath its supreme span a new secret is raging. A rising tide of young First Nations men and women train to mark a new history for their people, to bring new life to the wealth of their culture. They paddle their canoes undaunted by the cast of its shadow.

GIFT OF THE RAVEN
(Alano Edzerza)

RENOVATION 1972-1999

Guinness Money

When American Bridge/Surespan had cast their bid to renovate the Lions Gate Bridge in 1999, they had underestimated its tall history, spanning the First Narrows of Burrard Inlet for sixty years. Built to set Vancouver on the cutting edge of the world, its ill-gotten foundations gave it false ground lain over dark secrets. Never failing since 1938, its cost then just under six million dollars, it did not immediately reveal its cheap and idiosyncratic construction. Guinness money had primarily been put into the British Properties development as personal investment with indefinite longevity. The bridge, given the least financial consideration, had proven its inestimable value but with a definite lifespan. Eighty-six and a half million in 1999 was not sufficient to restore it to new life. Standing stalwart, it belied its sixty years, and hid the effect they had had. Also hidden was the neglect by governments for many of those years. If Buckland and Taylor Ltd., bridge engineers of North Vancouver, had not been hired to investigate the state of the bridge in 1972, it would not be standing. The story of the real renovation begins then and continues, giving Lions Gate secret life for at least one hundred years!

The dark secret, which wasn't revealed then or during any part of the renovation process, lay under the layers of power commanding the capital to be tendered. The First Nations' land from which Lions Gate had taken life and on which it was about to fall had not been remediated. The forced forfeiture of aboriginal homeland, the land prime to the lives of First Nations people would continue to bear witness to the ongoing ravages of colonization.

No Guinness money, paltry as that may have been, would be offered in remembrance of the wrong. And governments at all levels were never to admit to it.

The hundred years of extended life, given to the bridge by renovations, was nowhere accorded the First Nations people. Lions Gate would rise to new life. First Nations would return to the waters to fight for theirs, a secret they own.

The Seventies

The leaning towers of Lions Gate were the secret of the investigation. Originally designed without an extra safety margin, the towers would be in danger of severe bending stresses should the cables stretch or the saddles slip, which is what Buckland and Taylor found in 1972. The North Tower sitting on gravel had settled. Had it been left unsolved, the cable would have broken, the bridge fallen. Collars were fixed to prevent further slippage. To correct the lean on the North Tower, a brilliant invention, kept in place until November 20, 2001, was the secret.

A steel rocker bearing, inserted transversally ten feet into the concrete footing of the north cable bent, gave it freedom to rotate and the tower to straighten. Two steel plates bolted to move the bushing at the south end cable bent, forced the tower to straighten. The South Tower came back perfectly, the North Tower almost. Thereafter, the safety of the towers never wavered for thirty years. In 2001 when the security of the bridge was wholly assured—its weight having been rearranged to straighten the towers within millimetres—the rocker bearing was sealed off, much to inventor Peter Buckland's dismay!

The state of the North Viaduct, forty per cent of the total suspended structure, carrying twenty-five spans, could not be kept secret. The North Tower pier pedestals were found to have horizontal cracks that were later sealed with epoxy. The roadbed and sidewalks had been so severely corroded by salt accumulated throughout the cold wet seasons and worsened by warm rain running it into the rivets and rails of steel, that replacement of both was mandatory. Rust, the greatest enemy of steel, had overridden it on the seven-inch-thick roadway, creating one thousand potholes per year. In 1975, an orthotropic deck, lighter and fifty-seven per cent wider than the original concrete deck, was put down one thousand feet at a time during night closures. Sealed with epoxy paving, it yielded a lighter and slimmer design.

A great number of maintenance designs were ignored. Buckland & Taylor recommended the tightening of fifty-six cable bands by replacing three hundred and thirty-six bolts in 1977, and the cleaning of drains in 1979. These were not implemented. The most significant neglect of the seventies were the 'plans and specifications, ready for tender ... developed to replace the suspended structure of the bridge with three wider lanes, matching the lane widths on the 1975 deck replacement of the North Viaduct. This work did not proceed to construction.' (Buckland & Taylor Ltd. Bridge Engineering Report No.31)

By the end of the seventies, the North Viaduct replacement had completed the training for the future renovation of Lions Gate. Only once had the reopening of the bridge been delayed one hour. By 1979, Buckland and Taylor had realized the design for the major renovation.

The Eighties

Straightened towers and a new North Viaduct had enhanced the surety of the bridge. Still, the bridge remained susceptible to ships. While ships passing under rarely go off course, ten are known to have hit bridges in Vancouver. Lions Gate suffered one hit by a crane boom, giving it a ten per cent chance of being hit at any future time. When the North and South Tower pier pedestals were tested for ship vulnerability, the South pier pedestal was found vulnerable only to cruise ships. A protective concrete collar was anchored to the bedrock and steel rebounds were added for greater earthquake protection. One ship that lost course managed to steer between the cables and foundations of the South Tower, luckily missing its pier pedestal. The North Tower pier pedestal sits on a gravel foundation sixty feet deep that is subject to tides, making little ships vulnerable at high tide and big ships vulnerable at low. Medium ships are just the right weight to stay the course. A rock front added to the pier pedestal made it imperceptible to change or damage.

What remained perceptible were the towers, the main cables and the hangers. Preventive steps to public accessibility included fixing chain link fencing on the main cables, removing lower ladder rungs on the hangers, and installing a security system with alarms and emergency telephones on the towers, newly fitted with floodlight brackets that were not used. Unused designs by Buckland and Taylor continued to be shelved by the government. A deep drainage system, designed in 1983 for the south anchorage, was not implemented. Two decades of work had produced the surety of the two towers and their foundations—a vision in new engineering design—a sure bridge that could stand up to the stresses of the future but not without careful and complete renovation. The secret of the eighties? 'Since 1985, Buckland & Taylor Ltd. (had) studied options for renovation of the Lions Gate Bridge to carry four lanes instead of three.'(Buckland & Taylor Ltd. Bridge Engineering Report No. 31)

The nineties would bring that, surely.

Four-lane bridge design.
(Courtesy Buckland and Taylor Ltd.)

The Nineties

From the beginning of Buckland and Taylor's assessment and evaluation in the seventies, the challenges of Lions Gate had set standards and records in world bridge engineering design. Computer models had been non-existent and had to be created. Later generic models had not been sophisticated enough. Lions Gate had outmoded them. Cable software, written by Buckland and Taylor, is still the only software that models cables properly. By the late nineties, when the final design for renovation was being refined, the sleuthing job of three decades would result in a geometric solution unequalled in bridge history. The renovation itself would prove to be 'by far the most technically complicated job I'll see in my career', Darryl Matson, 1989 UBC graduate and project engineer for Buckland and Taylor, would declare. 'It's the first time it's been done. It's the first time that a complete hole would be cut out of a bridge, with nothing connecting it, and be ready for traffic in the morning! I am very pleased with the end product. They've done a top-notch job!'

*New bridge sections replacing old.
(Buckland and Taylor Ltd.)*

The decades of repair had improved the safe life of the bridge. Traffic loading, wind and turbulence had been tested and the bridge upgraded. But traffic volumes continued to increase. Pressures to the main span accumulated. Rust on the underside of the roadway deck, despite the emergency measures for extreme salt corrosion, persisted. 'Crevice corrosion has begun at nearly every crevice. It is many years since these were painted above deck.' The 1993 Lions Gate Bridge Study would warn: 'The most serious deterioration is occurring in the roadway deck, stringers, floor beams and sidewalks. This part of the bridge requires the most attention in the next five years.' (Buckland and Taylor Ltd. Engineering Report #20.) The paper-thin stringers fastening the deck were given a remaining safe life of two years, the sidewalks and teegrid deck, one and the wheel-track paving none. Supplemental beams were added as a temporary measure. Wind and rain and cars and trucks and bicycles continued to roar across the bridge. In 1992 the North Viaduct roadway had to be repaved. Replaced with a new, light orthotropic deck in 1973, it had stood the test of nineteen years with the original bond lasting twenty-six. In 1996 the viaduct sidewalks had to be replaced. Exuding the new slim design, the North Viaduct gleamed with new paint, while the bridge towers blistered with rust. And were those holes in the roadway or just the sun glancing off its shine?

Studies since the seventies had determined modifications for roadway loading but the rate of deterioration of the main roadway deck had increased. By 1995 the bridge carried twenty-five million vehicles per year, 'one of the highest vehicle counts per lane in the world', according to the Lions Gate Bridge Study. 'The ravages of traffic, the weather and de-icing salts continue unabated.' Leaks into the teegrids, fatigue cracks and breaks in the welds caused by rust had occurred in rapid progression. Patching patches had become an exceedingly expensive, very disruptive and extremely ineffective way to deal with the problem. 'Although the original structural integrity of the bridge has been upheld and even improved over its 57 years, deterioration of some components will continue to accelerate and cause ever-increasing maintenance costs and disruption unless major renovation is undertaken. It is recommended that reconstruction or replacement of the Lions Gate Bridge be completed within the next 4 years.' (Buckland and Taylor Ltd. Engineering Report #23) The cost of maintenance would be one to two million dollars per year. The sidewalks, stringers and deck expansion joints were given a remaining safe life of one year. There were no remaining safe life years in the driving surface despite the paving of wheel tracks and the paint system of the entire bridge. Rust blisters splattered the towers. The 1997 Lions Gate Bridge Study recorded that 'In June 1996, cracks ... in the webs of the stringers framing into the floor beams at the main towers ... opened by as much as 5mm when heavy traffic passed by on the deck above. Temporary emergency repairs were designed and installed quickly to avoid a partial collapse of the deck system.' The study confirmed that 'the remaining life of the floor system members is short ... It is recommended that reconstruction or replacement of the Lions Gate Bridge be completed within the next 2 years.' (Buckland and Taylor Ltd. Engineering Report #28) The original roadway deck had no useful life left. There was little time. The remaining safe life was being tendered by urgent repairs.

The 1998 Lions Gate Bridge Study set out 'to identify urgent repairs that may be needed to ensure the safe service of the bridge until at least 1999.' Reporting a 'faster rate of deterioration' in the top of the deck, 'the existing wearing surface, including the partial depth surfacing placed in 1993 is failing. Extensive repairs, or complete replacement, may be required in the very near future.' (Buckland and Taylor Ltd. Engineering Report #33) It recommended that 'rust spots on the outer surfaces of both legs should be cleaned of oxidized material and repainted to prevent further corrosion of the tower legs.' The tower brass bearing plates had full-depth cracks. Most nuts and bolts were more rusted than the rivets whose heads had long fallen. Paint chipped and peeled off the main cable. Stronger less corrosive materials with half the amount of steel in a design that would be much easier to repair would be functionally, visually and economically better. By the end of the nineties the lights in the towers had burnt out. Buckland and Taylor lit more of theirs and fine-tuned their final design.

RENOVATION 1999-2002

The Contract

The contract that American Bridge Canada Ltd. and Surespan General Contractors Corp. signed with the British Columbia Government on April 30, 1999, for renovation completion of the Lions Gate Bridge by December 30, 2000, included an earthquake upgrade, the bridge deck replacement, and a wind upgrade. Painting the bridge was not in the final contract. A provisional item, the government's decision to remove it was based on getting the maximum life out of the paint as with the roadway. The long-term maintenance by Buckland and Taylor had improved that life considerably. Their renovation had included painting as would American Bridge/Surespan's. Ironically, colour changes in painting the lower tower legs had resulted in added costs. Less than half of the area left to paint was hidden from salt and rain. Present environmental standards demanded that lead-based paint be removed using negative pressure (a vacuum effect). Individual painters would wear specially designed full suits, which would be thrown away each time the painters would re-enter the natural environment.

Ten million dollars, the estimated cost of paint removal and painting, is the cost of a small bridge! The painting of Lions Gate Bridge is the responsibility of the South Coast Region. Design engineer Peter Buckland said, 'It cannot be ignored. Aesthetically there is a huge demand. New steel demands high quality paint. Then it will likely last much longer. We did clean and paint in every part that we could get at. It should be painted every twenty years.' In truth it had not been painted for sixty.

Earthquake Design

While the Lions Gate Bridge was saved from falling down in the seventies and eighties, the Tacoma Narrows Bridge which had fallen in the 1940 earthquake and had been replaced was being assessed. Prior to 1940, earthquake standards considered only ten per cent of the structure. In 1984, Buckland and Taylor assessed the Tacoma Bridge using a powerful survey-and-analysis technique developed on Lions Gate. Both bridges were then brought up to modern standards.

The seismic upgrading on Lions Gate has continued to evolve. Design codes that Buckland and Taylor had tested on it for twenty years have been published in the Design Codes of Canada. They have had to be far ahead of conventional wisdom in most areas of engineering design and especially those pertaining to earthquakes. Their assessment in 1995, that a seismic retrofit was a priority, resulted in 1997 designs that were not used.

The criteria that Buckland and Taylor used to determine earthquake standards seemed simple: strength of the structure, so it doesn't break; ductility, so that it will bend and not snap when it's overloaded; tuning, so it can 'tune out' when the earthquake comes; movement, so it is flexible but doesn't respond to the dominant, fast frequencies of the quake; and damping, so it doesn't absorb the energy of the quake. During an earthquake, the bridge becomes resonant with the movement of the quake, moving first in one direction, then the other. The higher the bridge, the greater the resonance, hence the greater its chance of falling down. If, on the third cycle of movement the bridge cracks at a forty-five degree angle, the bridge falls. Lions Gate as a suspended structure has nine seconds of resonance, which has been synchronized with the fast frequencies. In its upgraded state, it has been 'tuned out' to earthquakes. Stability, ductility and more movement have been added to meet modern standards. The steel collar added to the North Tower pier pedestal in 1988 has been wrapped in fibre to give greater flexibility. At the cable bent, the bearing has been strengthened. The Buckland bearing installed in the North Tower pier pedestal in the seventies had to be fixed in place to lessen resonance. Deck braces have been added. At each end of the main span, the length of the bearings has been increased and the bearing shoes have been cantilevered to allow the span to travel during an earthquake. Four expansion joints in the roadway deck, two at the towers, one at the north cable bent, and one at the south end, are one and a half metres, giving greater flexibility. The North Viaduct, the shortest in height, making it the most vulnerable, has been worked on since the seventies. Complex but completely safe, Lions Gate Bridge has had the benefit of the most knowledge and best expertise in the realm of earthquake design, and is ready for the big one!

Roadway Deck

Custom-designed jacking traveller.
(Courtesy Buckland and Taylor Ltd.)

The width of the roadway deck has been increased forty-seven per cent, from nine feet eight inches to eleven feet eight inches, giving the two curb lanes the edge over the middle lane. Fifty-four new panels of orthotropic deck have replaced the old roadway. Choreographed within ten hours, fifteen stages executed the process during each night closure. Lowering the bridge, cutting through it, lowering out the old piece, lifting the new piece in place with the jacking traveller, bolting it to the roadway, raising the bridge to its height, and finally welding it to the new roadway composed the main part of the process.

Fifty-six transverse welds, equivalent to twenty-nine and a half miles, fixed the roadway deck in final place. Bolting, the most expensive, required eight hundred bolts per splice and forty thousand bolts just for the deck panels. All bolt and rivet holes, old and new, had to be filled with a bolt.

Hangers

*New hangers and widened sidewalks.
(Courtesy Buckland and Taylor Ltd.)*

Two hanger-length pieces weigh one hundred tons. Made of six steel strands, galvanized with zinc and set around a seventh core, each piece rises from the span (the orange marking its perfectly straight position), wraps around the main cable and returns to the span at a three-foot distance. A built-in jacking system tunes them, much like harp strings are tuned. Tuning the hangers means giving each equal weight. One hundred and sixty-six hangers, eighty-three on each side of the span, thirty feet apart, keep the weight of the bridge in perfect harmony to the main cable, which at thirteen inches in diameter carries it with the help of the two trusses. Thus the bridge is suspended. With forty per cent more usable space than the original, it is the identical length and weight, slimmer and stronger—like a swimmer. The new hangers can take five or six times what they could previously. Two can be removed in the case of a crash without disruption to traffic. Twelve traction rods, four mid-span, four at the cable bents, and four at the south end designed for earthquake flexibility by Buckland and Taylor, ensure a hard strength. Clamping added at the cable ends and over bolts to prevent sliding has given it more muscle. And secretly wrapped in the main cables, untouched cedar strips renew the bridge for another lifespan.

The Sidewalks

The original bridge sidewalks had been widened in the seventies from four feet three inches to six feet nine inches. Replacement of old sidewalks, which had begun in 1986, had been completed in 1997. The final design, in 1998, separated them from the main roadway with a thirty-three-inch-high galvanized steel girder. The secret to its design is that it minimizes crashes. A car out of control cannot go over or through it, crashing into pedestrians or cyclists or back into the oncoming traffic. Made of galvanized steel, it is lighter but stronger than concrete. The new sidewalk width of eight feet nine inches gives a two-metre clearance for pedestrians and bicyclists between the hangers and the railing. The fifty-five-inch railing meets the safety standard for cyclists and protects both. At approximately one million dollars per person, the view is worth it. To give it even greater value, Buckland and Taylor have added four viewing platforms to the new bridge design, two per tower, one on each side of the roadway. The view from them is spectacular! It is here that the design of the bridge is fully revealed. It brandishes its armour of new steel, braced for storm and torrent, shielding drains, funnels and lights that illuminate the towers resplendently at night. And 'the best kept secret' is the view from their portals, day or night. Long cold night shifts prompted Eduardo Pradilla, liaison engineer for the project, to transcend it all by climbing the tower staircase to see 'all of Burrard Inlet. The sunrise and sunset were stupendous from up there!'

Two-foot widths of additional sidewalk have been welded to the North Viaduct sidewalks to align with the new on either side of the main span, which have been repaved identically. The railings have been sprayed with green plastic coating.

Winds

Wind testing done in the seventies on Lions Gate had resulted in giving it stability in winds of one hundred and twenty-six kilometres per hour. Built originally as one of the longest bridges and the cheapest, no money had been spent on torsional stability. In the final design, Buckland and Taylor increased the torsional stability significantly to prevent the bridge from twisting. Beneath the roadway deck, an eight-foot layer of bracing can be seen from the viewing platforms. With additional vertical stability, the bridge has been upgraded to withstand winds of two hundred and seventy kilometres per hour. In 2001, general foreman on the Lions Gate Project Dean German would confirm that the bridge is safe in 'unpredictable wind. It's not going anywhere. It might scare a lot of people but it's not going anywhere!'

Canron

The eighty-six and a half million dollar contract with *American Bridge/ Surespan A Joint Venture,* signed in April, 1999, specified that British Columbia be the locale for fabrication. American Bridge/Surespan awarded Canron Construction Corp. West the twenty million dollar contract. In June of 1999, the fabrication schedule was agreed to with Canron beginning the complex process in July. By January 4, 2000, the first deck piece was out and ready, and the last by September 18, 2000. Within a three-shift, four/five day cycle, seventy-five crew including one woman, kept the fabrication on schedule, meeting the high standards of the design. Only the lugs securing the hangers had to be redesigned to be bolted vertically underneath rather than welded onto the deck.

The new roadway deck, designed to ward off extreme corrosion factors that the previous deck had shown to be the greatest cause of deterioration, was the biggest challenge. Planks beneath the old three-inch deck had broken and tiles were breaking. The stringers, steel longitudinal beams underneath, were paper thin by the time the renovation had begun. The challenge in the new deck design was to subvert the maintenance cost of three million dollars per year by strengthening it and adding width, to it and the sidewalks, without adding weight to the bridge. It comprised a five-eighths inch thick deck plate, with the floor beams and sidewalks welded in. Two matching panels would be assembled at one time—deck, trough plates, strut beams and sidewalk—a single piece all done in a shop specially designed for the Lions Gate Project, and all done upside down! Over one hundred miles of welding would fit the four pieces and the two deck plates together. Welding the trough plates, sub-fabricated in Portland Oregon, was the most difficult. Sub-arc welding has almost no room for error. The floor beams, pushed down onto the troughs creating tight tolerances, resulted in curvature. Shrinkage could cause misalignment. Holes would be drilled and aligned with custom-made splice plates, the final step in an assembly that took three and a half days and repairs just four hours. Clarke & Pattison, subcontracted to sandblast, paint and pre-pave the panels ready to be barged to the bridge site, worked in a tent connected by rails to the assembly site. The precision required was extraordinary and more sophisticated than the international standard. Accuracy for the entire fabrication process ranged from one half millimetre on the welding to a tolerance of five millimetres on trusses. The profile of the deck had to be exact longitudinally and transversally, meeting the contour of the bridge once the traffic barrier was attached. And it had to be safe. The welding, tested ultrasonically, proved it safe within three per cent of penetration. By the end of the fabrication process, Canron had set a new standard in fabrication design.

Gordon Ward Hall, vice-president of Canron, had been their operations manager in 1975 for the North Viaduct deck fabrication. 'In 1975, we expected the main span to be done in 1976. I was very disappointed that we had to wait twenty-four years before it was done.' Responsible for all erection schemes in 1999, he claims, 'The fabrication was demanding but the erection was the most complex I've ever seen. And I've been involved with a lot of bridges.'

American Bridge/Surespan
A Joint Venture

The contracting engineer had many problems to solve before the bridge could be reconstructed. The major one was designing the erection equipment. 'It proved to be a lot more of a problem than anyone had anticipated', assistant project manager Carson Carney would admit. 'It is an amazing feat that the engineers were able to come up with an alternative to the original concept. We had finished the engineering for erection the month before. It was a moveable truss ramp, a bridge within a bridge that traffic could go over, and wouldn't need many closures. But the debates about traffic loading, debate between the designers and owners about the requirement for bridge traffic, prevented it. We believe it could have worked. The executive decision was, "We need to go another route." We fast-tracked. A lot of risks were taken. The amount of time required to redesign added six months, causing the delay from September 1999 to March, 2000.'

An erection traveller was modified to lift and fit each new piece after cutting and lowering out the old, first lowering the bridge, then raising it with the new piece bolted in. Sixty-four feet long, weighing sixty tons, the equivalent of two fully loaded semi-trailers, it had four strand jacks in operation with two in reserve. Carney says, 'The main cable was the limitation. At thirteen inches in diameter, a normal traveller would have crushed it. The new traveller grabbed onto the hanger ropes.' Moving the jacking traveller took eight hours of a two-day cycle, the first for preparation, the second for deck replacement. A continuity link, used on the side spans, had to be redesigned for the main span. It took two and a half hours to disconnect the continuity link, one hour of cutting time, three hours lowering, repositioning and lifting, and three and a half bolting and getting the geometry of the bridge right. 'Changing the geometry of the bridge was a challenge. We would take a snap picture of the geometry of the section being replaced. Then, a trial fit. Then the most difficult task was to bring back the curve, the right road geometry. The new curve was shallower.' On alternate nights, welding would be the most difficult task. The thinness of the deck plate allowed for no movement, no error. Done in a tent with ceramic backing, 'it was a picky weld but state of the art!' That left half an hour to get the new bridge ready for traffic in the morning! Erection engineering of this magnitude, in ten hours! Fifty-four deck panels meant fifty-four different bridges, each needing the assurance of safety. The new roadway deck would not line up with the old to which it had to be connected. It was forty-seven per cent wider and much slimmer. The continuity link, designed to connect the new with the old, would transfer new sections from the ground, connect them in the air and ultimately to the bridge truss. Bolting and welding had to be done on sequential nights. The heat from the weld on the second night would keep back the paving. The transverse weld could only be done at night. Rain running down the roadway made it difficult. The trusses were levelled out of shape. It rained and stormed.

Ron Crockett developed a spring action on the continuity link to give cross-storm flexibility. As vice-president of American Bridge and its chief representative and senior engineer for the Lions Gate Project, he and Ugo del Costello were the heads of the new erection design. 'He's been on countless projects' Carson Carney emphasized. 'He's very intelligent.' Coming directly from renovating the oldest suspension bridge in the world in West Virginia, Ron Crockett's experience lay in renovating and refurbishing old suspension bridges rather than making them new. A 'first of its kind' reconstruction project in Lisbon Portugal, where a bridge was widened from three lanes to six and a railway bridge added, all under live traffic, had been his recent challenge. Lions Gate would prove the greatest challenge. 'From an engineering point of view, it was much more complicated than was apparent at first. It took eighty thousand man hours of engineering, that's forty man years, to figure out the engineering! And twenty-eight months to do! It had never been done in this way before. Keeping the bridge open to traffic on such a grand scale had never been done. The engineering effort wouldn't have been as great if we had closed the bridge. It was impossible to close. It would have taken a full month of full closure. It would have been devastating. I think they made the right decision.'

The extreme end of the south side span did not have access to it underneath the bridge. A more sophisticated jacking traveller had to be designed. Old deck pieces had to be cut out in ten-metre lengths, lifted onto a turntable, rotated, and skidded across the gap to be transported off the end. The new pieces, brought in from the north end, had to be skidded across the gap, turned and lowered into place. A crane would be required to replace the last two sections. American Bridge/Surespan engineer Kevin Smith coordinated all engineering activities.

Assistant project manager Carson Carney says 'The wind caused the biggest problems. When it blew, the bridge moved around like a snake. We saw half of what it was designed for. During the biggest storm, we saw two thirds of what it would have taken! We consulted Environment Canada for wind studies. They were able to predict down to the metre per second; to fifteen metres per second, to ten metres per second, right down to eight. We had a hand-held metre and monitored it with a report from them at four p.m. to see if we could go. We never had to cancel any lower-lift nights.' Winds in December 'would have taken it, had we been working on it. We were lucky.'

Hard work rather than luck made the first closure the longest, at seventeen and a half hours; the next thirteen hours, then twelve and a half, then ten, and the best time, eight hours!

The first piece of the roadway deck went in September 9-11, 2000, and the last, September 29-30, 2001. Of the old bridge, only the cables and towers remain. The bump, the unremitting evidence of Guinness cheapness, had gone! The stiffening trusses welded to the deck plates had yielded a slim smooth line. Peter Buckland thinks, 'It's better than I thought it would be. I did not expect the visual effect to be so great!'

Ron Crockett believes 'it is going to be quite a landmark for years to come. Most of the suspension bridges don't have that open feeling. Usually you can't really walk and see or drive and see everything. The way the truss is designed under the bridge! It is beautiful. And such a grand view!'

Mediated Settlement

But there were still a few bumps on the road. One year behind schedule, the contract carried penalties for extended closures and delays. In January, 2000, eight million dollars for the Stanley Park causeway widening had been added to the original contract, now worth 108 million dollars with the extra project costs. In November, 2001, the British Columbia Ministry of Transportation and the American Bridge Board of Directors agreed to a mediated settlement of seven and a half million dollars going to *American Bridge/Surespan A Joint Venture*. The settlement included over one hundred claimed extras and over one hundred work orders. An additional ten million would take care of unanticipated technical, safety, environmental and management costs, as well as a half million dollar contingency, the first and last such allocation. The final completion date of December 30, 2000 was changed to June 30, 2002. The total delay of the renovation would be equivalent to the total building time of the first Lions Gate Bridge. The renovation, costing one hundred and twenty-five million dollars, would be about twenty times the cost of the original. Senior engineer Ron Crockett sums it up as 'certainly the most difficult project, more complicated than any other. It was the most gruelling experience, with all the pressures the contract placed on us, the most challenging and the most aggravating. But because of the tremendous effort, we could put eighty thousand man hours of engineering into six months. When we got into it, it was a lot. And it didn't cost that much more considering the grand scale of the project. We've been building bridges since 1900. There are a handful of contractors, four or five in the world, who can handle these projects. We were the low bidder. We got the job. In spite of the fact that it took so long, it's very rewarding.' Assistant project manager Carson Carney, a 1995 graduate of Carnegie Mellon University, had come directly from Providence, Rhode Island, where a mall complex in the middle of the city, complete with bridges, had taken one year and four months of his time. 'Lions Gate Bridge is ten times anything that I've ever seen. It has been such a large accomplishment recognized throughout the world as an amazing feat! I can't say I'd ever want to do it again! There've been a lot of grey hairs and ulcers because of all the problems. I would like to avoid them in future. I'd love to build a new bridge. It's tough going place to place. My wife has been amazing! She's supported me throughout. I couldn't have done it if she hadn't been up here. I could not do that. But she is in limbo.' By the end of the Lions Gate Bridge Project, Kelly Clements Carney would have her own achievement: a Master's Degree in Educational Psychology, done online through an American university.

Secret Power

The final paving, at a cost of 1.5 million dollars, is high-tech epoxy that possesses secret bonding power, possible only at high temperatures. Orchestration is key in the paving process. Timing and temperature need to be perfectly in tune. Moisture free air and steel temperature have to be ten degrees Celsius and rising. Shipped at one hundred and forty degrees in trucks, ten at one time, it cannot be poured sooner than forty-two minutes and not later than sixty-two minutes after it has been mixed. Otherwise it cures and has to be dumped by removing the truck panels because it has solidified. For it to be flexible and durable, which the roadway deck demands, temperature, timing and thinness are precision factors in this weather sensitive process. The epoxy bond is a half-inch thick and the paving one inch for a precise layer of one and a half inches, not one millimetre more or less, for a perfect weight of three hundred and seventy pounds per foot, equivalent to a line of bridge traffic! Left on the job alone, Carson Carney exuded patience. 'We need one good weekend and a dry week preceding it. That usually happens in July and August. And that's tourist season! We're hoping to get it in before the end of June.' As Carney predicted, July came hot and dry. Close study of its weather patterns led to the decision. The process was orchestrated during one night's closure, July 20-21, 2002.

Liaison engineer Eduardo Pradilla admits that risk is worth it. 'It was a gamble. We had to know by Wednesday when BA Blacktop could begin heating the mix for paving Saturday. There was a ten per cent chance of rain. We decided to go for it!' He confides that he had summoned higher power. 'On September 9, 2000, we had just laid the temporary surface. In my village in Spain they were celebrating the Festival de la Virgen de la Antigua. I telephoned my father to ask him to light a candle for me.' Almost two years later, the temporary surface was intact and cleaned with pressurized detergent, rinsed and cleaned again, swept and vacuumed during the two night closures preceding the final paving. Sub-contractor BA Blacktop Ltd., 'the same company that paved Golden Gate with the same stuff' was finally paving Lions Gate. 'Very sophisticated and very expensive!' Peter Buckland concedes.

Fifty-four men were summoned to the task. They came with extensive skill, experience, and their own secret power. In just ten hours they would conduct the immense work with a rhythm that belied its intensity. Six hundred tonnes of mix would be dumped from synchronized trucks, arriving every five to six minutes, into a temperature-controlled paving machine that poured it in perfect layers onto epoxy bond, spread manually two hours ahead for perfect adhesion. One hundred degrees to start, exact temperatures, read with infrared guns at each stage, are critical. For long life, compaction must be completed before the temperature drops to sixty-six degrees. Three compactors, two with rotating steel wheels and one with pressurized tires, ensure that longevity with a back and forth synchronicity at astonishing angles until the excess is sealed in, pebbles removed and all air released for greatest absorption. Timing and temperature are the urgent criteria in this race. Veteran drivers turn it into music.

Men with steel brooms work the joints between the lanes, carefully, closely, specially, to exact a perfect seam. One man, turning the mix with a long steel rod before it is released, keeps pace with the paving machine. One other constantly cleans the compacting wheel, running with it backwards, forwards, trowelling and stamping patches missed, and back again, not missing a beat. Another runs with a wheelbarrow to add or remove the hot mix wherever needed. The process is as unstoppable as the men, and heavily manual. The expansion joints are filled flush by hand. Those in closest proximity to chemical toxins wear protective gear, full respirator masks, and special coveralls. The man who swings the epoxy rod in a rhythmic arc, spraying bond continuously, has to be strong. They all must be. Helmeted and hungry they fight the elements, natural and toxic. They do not stop until it's done.

It didn't rain. For those who decried the two-hour delay, the bridge reopened with hand-marked lanes, black and beautiful. 'For every hour of work, there have been ten hours of planning,' Eduardo Pradilla contends. 'For a ten hour night there have been one hundred hours of planning to ensure that it be as smooth as possible. These men deserve our admiration. As much as technology advances, people are still the best. In a good worker, the human component is very important. All the bridge workers have impressed me. They have been professional from first to last.'

And the last word from one of the professionals, who has been there from the first to the last? 'A huge weight has been released from everybody's mind. It's going to be a good bridge.' The final test proves that. The profile of the finished bridge matches perfectly with the design.

For architect Barry Griblin, whose proposed design had received study, 'the concept of retaining three lanes is ridiculous. But the design is spectacular!'

Safety

No one had been killed! Safety, the largest requirement during construction and reconstruction, had been Buckland and Taylor's primary component as the independent review body. Proposed contract methods and the finished products had been reviewed. The quality of construction and activity had to be confirmed. Nothing could be sacrificed or compromised in safety and design. They were out on the bridge most of the time. They were in India to review the manufacturing of the hangers. They have been responsible for quality control and quality assurance that the bridge is operating properly and safely. Over twelve thousand documents are proof of their vigilance.

It is no secret that ironworkers are killed on a regular basis. Contractors make money by taking risks. One in two deaths are a result of too few regulatory boards or too many, not because of bridge failure. One in ten thousand deaths are due to bridge failure. For the Lions Gate renovation, three bodies have met often and have had to agree before work could continue: the BC Transportation and Finance Authority, the financiers and directors for the project; American Bridge/Surespan, the contractors responsible for all construction methods; and Buckland and Taylor, the designers of the project, who provided the independent review. All three had to reach agreement each time quality or safety were at issue. All work had to satisfy the Bridge Code.

There were no structural accidents. One equipment-related accident, involving a cable buggy that spun out of control but stopped before anyone was killed, resulted in two workers being injured. Another suffered a broken ankle in a separate accident. Neither accident interrupted traffic. Buckland and Taylor's role as the independent-review body continued to the end of the contract. They have been the owner's bridge engineer since 1972.

Accident prevention has been the priority of Nicolette Wilson, head of health and safety. As American Bridge/Surespan's representative, she had to coordinate everyone with everything, giving each new ironworker the safety training required for each procedure. At peak work times that had meant making sure that one hundred and ten workers and the public were safe. Beginning in August of 1999, it would take over four hundred orientations to assure that safety to everyone. Her responsibility was to 'have the job completed without injuries and to see that my company did not get into any trouble.' While written procedures within union regulations are available, the work on Lions Gate was 'the leading edge. There were no in-betweens, only extreme risks and near misses.' The main safety concern was fault protection: that workers were tying off their harnesses, that they were wearing their retractables, and that the specialized equipment and tools were tied down. Huge risks were endemic to the job. 'Imagine us taking a piece of the bridge out. If that piece fell, it would take down the workers on it. The effect on the bridge would bring other workers down. And the public! Everything these engineers did, putting in all these safety factors, would mean nothing. If anything happened, it would be a major catastrophe.'

To prevent any chance of catastrophe on her first bridge, Nicolette Wilson often had twelve-hour shifts. Routine night inspections called her to the site at three a.m., lengthening her day that began at five thirty a.m.. Nights were often plagued with dreams. 'You take it home with you. Until I see every worker leave here, then the job is still continuing. They have to understand, and they don't necessarily listen. These guys definitely want to go home every night. I'd rather prevent injuries than have to take care of them. Every single day I've learned something. I think I've found where I want to be.' Being one of few women on the bridge, she commanded the respect of the men. 'I have been proud watching them.' The disdain of the drivers has been her only discouragement. 'They were totally ignorant! They threw handfuls of pennies at the workers, shouting, "The government is wasting your money!"'

Monument to the Ironworkers

Eduardo Pradilla, the link for Buckland and Taylor to American Bridge/Surespan, graduated from the University of Santander in Spain in 1994 and has worked internationally on both sides of the bridge experience, constructing and contracting. 'As an engineer, I am thinking about the ironworker all the time. I try to make it easy, but that doesn't happen all the time. There is only one Lions Gate, its construction unique. It's an old bridge with elements of unknowns. A rusty pin being removed for the first time in sixty years! The ironworker improvises, finds his own method. You can't make a mistake! This bridge is a monument to the ironworkers working on this bridge. It is really the ironworkers who have been there through rain, storm and wind, twelve hours without a break, working as hard as they could. They are the guys doing the work.' Women too. 'They have great experience with steel work. They have saved the contractor time and money working long shifts. They are very proud to have worked on the bridge. They chose to stay. They wanted to do it in the best possible way. It shows in the quality. There have been over a hundred closures! The pressure on the night superintendent has been incredible. They did perform. It has been a success. It is slightly over budget but not that much considering the job. And it has been done safely. The noise has not been a problem. There have been very few complaints. The drivers have been the problem. Insulting the workers! But the e-mail support has been great.'

Dean German, general foreman on the project, worked nonstop shifts. 'Learning the new procedures and meeting the expectations of the company' were his daily challenges. 'They spring things on you constantly. It's been going on throughout. Trying to bring it down to a schedule that was manageable in order to get it done on time' was a singular achievement. German's secret was to 'shorten the schedule and increase productivity. The ironworkers were excellent. It could never have been done without their cooperation. They did everything they were asked to do, everything and more. There was always teamwork. In order to fulfil the schedule, everyone had to perform as a whole. I just made sure it was running smoothly and safely. That only happens with cooperation. I had to answer to everybody about everything! There is nothing comparable to the work on Lions Gate Bridge at this time. It's because of the crew.'

The ironworkers, all Canadian including Lynn Therault, were hired by the contractor from Highways Construction Ltd., which supplies the union. The night superintendent Harry Noble came from Britain. American Bridge/Surespan Assistant Project Manager Carson Carney says of the ironworkers, 'That's what gives a sense of pride. It's been the absolute toughest thing ever done and they actually did it!'

For Eduardo Pradilla, who had watched over everything that happened on the site, seeing it from both sides at the same time, 'the most important thing is the personal relationship. It is very important to interact with everyone.' Coming directly from contracting on the Jiangyin Yangtze River Bridge, the world's fourth longest suspension bridge in 1999, he says 'Lions Gate is internationally a big imperial bridge. I like travelling. I like bridges and big bridges don't happen in the same place. You have to follow them.'

Hazel Keist, one of the eight operating engineers responsible for manning the hoist at the north end, which brought the ironworkers up for day shifts at 6:30 a.m. or nights at 7:30 p.m., had been trained by her husband Gary Keist, who had worked it from October, 1999 to June, 2000. When he had become too ill with throat cancer, she took the job in June of 2000 and performed it until it finished on December 10, 2001. Gary Keist died in November, 2000. As a member of the International Operating Engineers Local 115, twelve-hour days did not deter her. 'It is the best job that I've done. Everybody's been terrific! This bridge will have special meaning to me, that I've been a part of it.'

Project Patrol

Capilano Highway Services supervisor for patrol and communications Dave Howard's part in all the years of renovation are noted with 'a sense of pride. This project was the biggest. It was a huge engineering job that's never been done in the world. I think they're going to get a lot of awards for this one. The contractors were great to work with, very professional and meticulous. Despite the length of time it took they did a great job.'

The Ministry of Transportation has upgraded the traffic-control system. 'This is a very difficult traffic management area because of the volumes of traffic coming from West Vancouver into the downtown core and the Whistler corridor adding to it on Fridays, Saturdays and Sundays.' Ferry traffic is an unpredictable constant. 'Our main concern is to watch the traffic to make it flow better. The only difference with the new bridge is that we have better exposure to the areas with more cameras ready for instant response. We know the problem sooner than ever before.'

Twenty-four cameras on the light bridges bring the big picture onto ten television monitors in the traffic-control room. Six pick up all the transit areas, four pictures per screen. Two zoom in or out for a more defined look. Two show the whole lane arrangements and can be used for segmented lane control. 'When someone is in distress, we can zoom in close enough to read the license plates. The cameras do three hundred and sixty degrees.' Bryan Griffiths, communications operator, is happy to be in improved conditions rather than the temporary trailer that had housed the traffic-control system for seven years. His career, first begun in communications, with an eleven-year interval in traffic patrol, resumed in communications, giving him greater perspective. 'If there's a snowstorm, trees down, and an accident, all happening at the same time, that's when it helps to know the job. We take all the information, sort it and give it to those who need it. The information we gather is from the Ironworkers Memorial Second Narrows Crossing to First Narrows and Highway One to Horseshoe Bay. We are paid to monitor and assist the efficient flow of traffic. All the monitoring system is designed for Lions Gate.' Enhanced for greater efficiency, it has drawn interest internationally.

The cameras try to catch the jumpers. Bryan Griffiths' worst experience while on patrol was finding a person in her thirties who had jumped off the bridge and landed in the park. 'It took a long time to find her. She died about fourteen minutes later. Now, if someone is on the bridge, you can zoom in.'

Patrol officer Sam Davies saved one person in October. 'I held her. Over the years I've done that a few times. Now we can see it, which is a lot better. We can respond as soon as we're dealing with a jumper.' Responsible for coning and deconing the lanes for all the lane changes, Sam Davies is one of the traffic personnel who wait in the truck at the Georgia Street entrance. Between emergencies, he fills his time drawing faces in pencil, pencil crayon and India ink. On point duty, he has seen many accidents at the Georgia Street approach. 'When they come seventy miles an hour at night, they can't make the turn. They're going to straighten it. It has two sharp corners.' When drivers crash at the sharp turn, they often end up on the centre girder or in the lagoon, like one motorcyclist Davies had to rescue.

Drivers who did not slow down for the bumps over the expansion joints during the reconstruction of the bridge became airborne. Those who did not observe the lights caused head-on collisions. Most lane blockages are attributed to driver error. Sam Davies patrolled Lions Gate for ten years. 'We assist the public. Flat tires, accidents, people running out of gas. I've had to shut it down because of trees coming down. Anything that stops traffic I deal with.' His job had not changed except for 'bus loads and bus loads of people taking pictures of the bridge. 'My truck is in thousands of those photos!' Of the reconstruction he says, 'We worked very well together and seemed to pull it all together. I think everybody was expecting it to be bigger. I think it disappointed people who were looking for another lane. They should have another span right here.'

Government

When the BCTFA funded the Lions Gate Bridge Project in 1999, they had not allowed for contingency. American Bridge/Surespan had won the contract on pass/fail criteria, the low bid and the fewest closures being significant. 'The absolute must was experience in this type of bridge work. It was most critical', says Peter Hyslop, the owner's engineer. 'The biggest problem for the contractor was that it was a unique job. It had never been done. Things occurred that were unexpected. Delays were caused because of the unexpected nature of things that had never been done before. There were unforeseen conditions with anything existing, usually underground or up in the sky where the cables are. We had that as an understanding. We had known that there would be some but the extent was not known.' Unanticipated conditions became extras that were paid for.

Peter Hyslop stresses the 'unforeseen unique conditions. It was really unforeseen when the experts in aerodynamic analysis wouldn't agree! There were PhDs on both sides with different opinions! When there was doubt, we had mediation.'

In 2001, the British Columbia Liberal government shut down the BCTFA. Its project director Geoff Freer moved on. For its final year, the Project would be run by the Ministry of Transportation. Mike Proudfoot, director of design and construction, with the project since the year-long proposal preparation in 1997, calls it a 'big accomplishment, a big commitment. It was the working relationship of these professionals that makes it such an outstanding project. It is unfortunate that it has taken longer than intended but when you look at the complexities involved in all those issues ….'

Proudfoot attributes the resolution to those issues to 'all of the experts from Buckland and Taylor and American Bridge/Surespan sitting down and putting their energies to resolving them. It meant a team of specialists working around the table, long hours. There were electrical specialists, traffic specialists, communications experts, the bridge experts, seismic… To have the opportunity to be involved with and surrounded by a group like that was quite a reward. It was a local, national and international consulting committee.' World-class feats were performed there and everywhere else. 'My appreciation for the work is in the field.

It is not only the engineers that have had to figure it out. But then to watch the workers with their blow torches hanging from the railing' was amazing. 'It's been an incredible journey. The cooperation from the North Shore municipalities, from the participants... the patience and understanding of the public has made this project a success. We had no alternative to closing the bridge. The public saw the benefit and responded accordingly.'

The immensity of the work at all levels has yielded immense benefits. 'Driving over that bridge now with panoramic vistas in all directions is truly remarkable. The facilities for cyclists and the viewing platforms give the public benefits beyond the wider, safer bridge.' Being on the bridge, Proudfoot exclaims, is to experience the 'alignment of the planets! What a perfect representation of this city, the most beautiful city in the world!'

Indian Land Claims

Beneath the spectacular sights, the project site, located on the BC Railway right-of-way, did not infringe on Squamish lands. 'It's out of the cut-off lands,' Gibby Jacob of the Squamish Band Council confirmed. For the first time, no changes to Capilano Reserve No. 5 resulted from major bridge renovation. Indian land claims for previous bridge infringements have not been processed. 'Lions Gate Bridge was not in the ninety-two and a half million dollar settlement we got in 1999.' No more secrets.

Guardians of the Gate

The great lions sculpted by Charles Marega at the Stanley Park entrance, who partially lost their right ears to the roar of debate and renovation, received the attention they command. They were sandblasted and power washed to their original colouring, their broken ears restored. Made new, in their individual uniqueness, they give added enchantment to the bridge. There is no telling what secrets they hold.

EAGLE LANDING (Alano Edzerza)

ACKNOWLEDGEMENTS

The delay in bringing this manuscript to fruition is not due to any reticence on the part of the participants. On the contrary, they have been the beacons that shone the way and revealed the means to get there. Peter Buckland's unwavering mastery of and loyalty to the bridge project inspired the urgency in the telling of the story. Thanks to him and his company of engineers, the innovative engineering secrets that the world had not seen, are unveiled. Eduardo Pradilla crowned the last morning by lending me a hard hat, checking my boots and accompanying me to the finish line. It was dawn.

Up to that point, the on-site co-operation was fantastic. Ken Crockett, Carson Carney, and Nicolette Wilson gave credence to the remarkability of the project. Dark secrets that emerged during the research process commanded their telling.

I am indebted to the First Nations for giving me a glimpse of the undue process by which they had been eliminated from the story. Thanks to Chief Gibby Jacob and Chief Bill Williams of Skwxwu7mesh for providing that integrity. Lisa Wilcox shone her light on their path. Without Bernadine Mathias and Mary Jacob to point the way, I would have missed it altogether.

Lilia D'Acres

ABOUT THE AUTHOR

Lilia D'Acres is an award winning author of *Lions Gate*. She has taught Writing and Literature committing herself to the literary arts. She founded the George Woodcock Centre for the Arts and Intellectual Freedom Fund, endowed at the UBC Library.

SOURCES

BCTFA Project 22. Lions Gate Bridge Project. Bridge No. 1481. Part 3. 1998.

BCTFA Project 22. Lions Gate Bridge No. 1481 General Information Present Bridge. Buckland & Taylor Ltd. Bridge Engineering A-F. 1998.

BCTFA Project 22. Lions Gate Bridge No. 1481. 3-Lane Renovation. Assumed Erection Scheme for Analysis. Draft Report #34. 19 October 1998.

Buckland, Peter G. Design Engineer. North Vancouver: Buckland & Taylor Ltd. Bridge Engineering. Interviews. 21 Nov. 2001 , 28 Nov. 2001, 5 Dec. 2001, 21 July 2002.

CSCE Sixth International Conference on Short and Medium Span Bridges. 2 August 2002, Vancouver.

Carney, Carson. Assistant Project Director. West Vancouver: American Bridge/Surespan. Interview. 13 December 2001.

Chan, Philip. Marketing Communications Manager, Buckland & Taylor. 21 April 2014.

Crockett, Ron. Senior Engineer. American Bridge/Surespan. Telephone Interview. 2 January 2002.

Davies, Sam. Patrol Officer. West Vancouver: Capilano Highway Services. Interview. 6 December 2001.

German, Dean. General Foreman. West Vancouver: Lions Gate Project. Interview. 18 December 2001.

Griffiths, Brian. Communications Officer. West Vancouver: Capilano Highway Services. Interview. 6 December 2001.

Howard, Dave. Supervisor Patrol and Communications. West Vancouver: Capilano Highway Services. Interview. 6 December 2001.

Howie, Carol. Administrative Services Division, West Vancouver Archives. 12 June 2014.

Hyslop, Peter J. Engineer. Vancouver: N. D. Lea Consulting Ltd. Interview 12 April 2002.

Jacob, Chief Gibby. Executive Operating Officer, Intergovernmental Relations, Natural Resources & Revenue, Squamish Nation. Interview. 01 May 2002.

Jacob, Mary. Administrative Support, Intergovernmental Relations, Squamish Nation. Interview. 26 May 2014.

Mathias, Bernadine. Administrative Support, Intergovernmental Relations, Squamish Nation. 26 May 2014.

Matson, Daryl D. Engineer. North Vancouver: Buckland & Taylor Ltd. Bridge Engineering. Interview. 26 November 2001.

Ministry of Transportation & Highways. Lions Gate Bridge Study Report #20. Bridge Condition Survey. 30 March 1993.

Ministry of Transportation & Highways. Lions Gate Bridge Study Report #23. Bridge Condition Survey. 31 March 1995.

Ministry of Transportation & Highways. Lions Gate Bridge Study Report #28. Bridge Condition Survey. 31 March 1997.

Ministry of Transportation & Highways. Lions Gate Bridge Study Report #33. Bridge Condition Survey. 31 March 1998.

Ministry of Transportation & Highways. Lions Gate Bridge Study Report #31. Summary of Conceptual Design for 4-Lane. Flat Cable – Stayed Conversion. 31 March 1998.

Pradilla, Eduardo F. Engineer. North Vancouver: Buckland & Taylor Ltd. Bridge Engineering. Interview. 4 December 2001.

Pradilla, Eduardo F. Engineer. North Vancouver: Buckland & Taylor Ltd. Bridge Engineering. Interview. 21 July 2000.

Proudfoot, Mike. Director, Design and Construction. Vancouver: British Columbia Ministry of Transportation. Interview. 19 December 2001.

Wardhall, Gordon. Engineer. Vancouver: Canron. Interview. 10 December 2001.

Wilcox, Lisa. Senior Executive Assistant to Chief Gibby Jacob, Intergovernmental Relations, Squamish Nation. Telephone Interview. 27 May 2014.

Williams, Chief Bill. Lead Negotiator Aboriginal Rights & Title, Intergovernmental Relations, Squamish Nation. Interviews. 28 May 2014, 11 June 2014, 12 June 2015.

Wilson, Nicolette. Health and Safety Representative. West Vancouver: American Bridge/Surespan. Interview 18 December 2001.

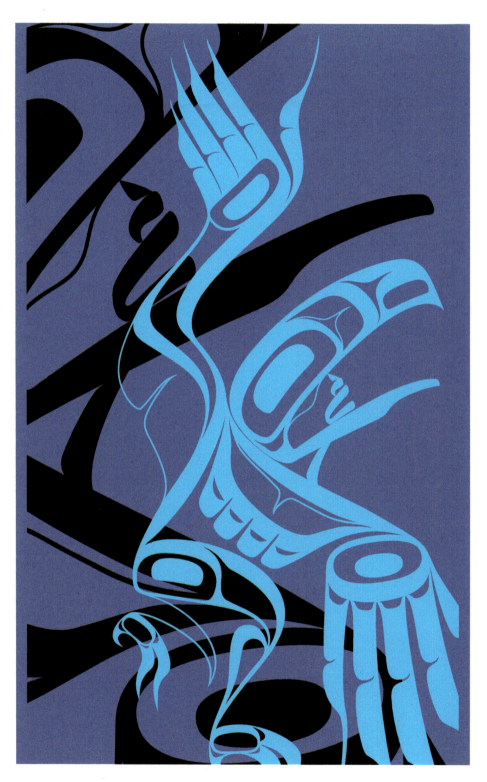

RAVEN DANCER (Alano Edzerza)

Copithorne
March 12 '17
Vancouver